THE VALUE OF WEEDS

Bulbous Buttercup or Crowfoot

TIIRD EDITION 1976

ISBN 0 905200 02 0

CONTENTS

Research by Joy Griffith-Jones,
The Soil Association

Bibliography

Book of Herbs by Dorothy Hall
Gardeners Earth by Stanley B. Whitehead
Gardening with Compost by F.C. King
How to enjoy your weeds by Audrey Wynne Hatfield
The Weed Problem by F.C. King
Soil Association Journal
'Mother Earth' July 1953
'Mother Earth' January 1954

THE VALUE OF WEEDS

Introduction

This is a booklet in praise of weeds. It has been written to help all gardeners to look upon their least favoured plants with more tolerant eyes, and when they see a weed to ask themselves, "Why is it there. What can it tell me and how can I use it?"

"WEED" says the Concise Oxford Dictionary, is a "wild herb springing where it is not wanted." A nice definition and most people would agree although few would recognise weeds as 'herbs'. Nevertheless, the word 'weed' has been corrupted from the Anglo-Saxon word 'weod', meaning herb or small plant.

From earliest childhood, most of us have been taught to exterminate, quite ruthlessly, every weed we see in our gardens. Not from a natural hatred of that particular plant, but simply because it is uninvited and "springing where it is not wanted" There is usually some reason why weeds appear in a specific place and the observant gardener should be able to recognise these signs. Once you come to terms with the fact that every garden will always have weeds – if the soil is at all fertile – and that there is a great deal to be learned from observing them, the problem can be approached in quite a different and indeed, in a much more positive way.

Chickweed

How do they come? Their seeds arrive in a variety of ways. Birds bring them; the wind carries them; they stick to our shoes and cling to our clothing and use many other subtle methods to ensure their survival. They are some of the most persistent and determined plants there are, and in many cases the seeds lie dormant until conditions are just right for them to germinate. Soil that has been dug or disturbed in some way, reveals and brings to the surface weed seeds that may have lain dormant for many years.

Perennials multiply both by seed and by vegetative reproduction. Most of the annuals multiply by seed but some also propagate vegetatively. For example, the seeds of the common chickweed, one of the commonest and most persistent of weeds, will germinate at any time of the growing season, and when its stems are cut or severed by the lawn mower, each small piece is capable of rooting afresh and making new plants.

There must be some rational explanation why weeds still persist in springing up, time after time, without any encouragement, and in spite of all the efforts made by man and his machines and chemicals to destroy them; while other plants have to struggle to survive in spite of all the skills and cosseting that man gives them. Ralph Waldo Emerson said: "What is a weed? A plant whose virtues have not yet been discovered." Let us take a closer look at some of our weeds and find out more about their virtues and uses.

Soil Fertility

Always remember that a weed is a green plant, with all the characteristics and qualities of such a plant. The ability to utilise the sunlight, to provide food for animal and man, and to fulfil the vital function of feeding the soil population. When the soil was being formed, a process that took billions of years, everything that had once lived was returned to the soil. It was by this cycle of life, by following the very important 'law of return' that the fertility was built up and maintained in the soil. In the top six or nine inches of a fertile soil, in which the plant roots mainly grow, is an enormous soil population working and living in symbiotic harmony with the plant life above. These workers are essential in helping to maintain the fertility of the soil because they provide food for the growing plant, which in turn provides food for both animal and man. Following the 'law of return', plants, animals and man provide food for the bacteria in the soil. This cycle of life, this natural decomposition of all animal and vegetable wastes, takes time, and so man speeds up the process by making compost. And there is no better contribution to soil fertility than an abundance and variety of weeds in the compost heap.

Small Nettle

If plants, which have appeared naturally in an area, are constantly removed and not returned in some form, the fertility of the soil will suffer because these same plants have been instrumental in building up and maintaining the fertility of that area. The appearance of certain plants, or weeds, is very often significant in pointing to some imbalance in the soil. The presence of weeds may indicate poor drainage, lack of aeration, or some deficiency, but whatever the reason, they should be returned to the soil, either via the compost heap or by hoeing and leaving them to decay on the ground. Weeds are usually plants which have adapted themselves to poor soil conditions and their flourishing may be an indication that the soil is deficient in some essential.

Many weeds are deep rooting plants which penetrate to the sub-soil and bring to the surface valuable elements not available to the shallower rooting plants. The decomposition of their leaves and stem enriches the soil, returning the trace elements and minerals taken up by the plant and adding fibrous matter. Weeds contain trace elements of varying amounts, some may be rich in one particular element while others may contain only a trace. But whatever the quantity, all the nutrients have been taken from the soil. If there is a good reason why the weeds should not be hoed and left to decompose in situ, then the natural and logical way to return these nutrients to the soil is via the compost heap. In this form they are much more readily available and acceptable to the growing plant than any of the chemical substitutes.

As well as being a storehouse of nutrients, those weeds which have long tap roots, or which are otherwise especially deep rooting, act as sub-soilers, aerating the soil and helping with drainage. How many times have you seen coltsfoot and dandelion thriving in heavy clay, the kind of soil that is sometimes difficult to put a spade into! These are two of the best sub-soilers; coltsfoot with its penetrating mass of creeping rootstock, and the dandelion with its long tap root. When the ground has been sufficiently aerated and the fertility built up, they will tend to disappear only to reappear elsewhere where conditions warrant the same service.

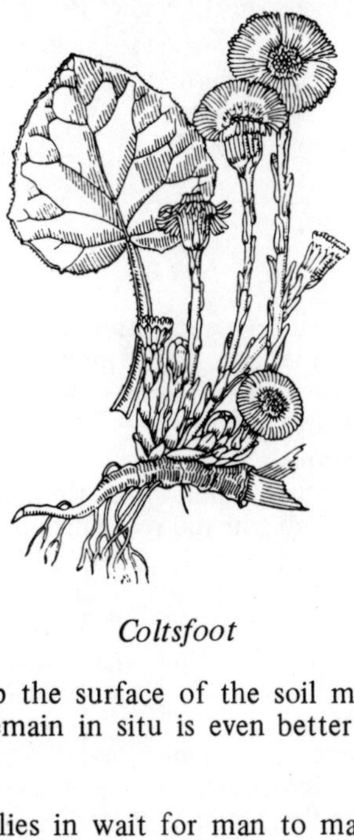

Coltsfoot

Weeds are capable of lessening the impact of heavy rain upon the ground, and provide shelter and food for small animals whose excreta help to fertilise the soil. In the hot weather weeds provide shade and cover from the sun and help to keep the surface of the soil moist. Hoeing them and allowing them to remain in situ is even better and helps to conserve the moisture.

Weeds as indicators

Somebody once said that the soil lies in wait for man to make a mistake, but it is more often the weeds that lie in wait, because whenever they appear they tell a story. The type of weed and where it grows can indicate lack of humus, drought, mineral deficiency, lack of aeration and so on. The wise gardener would do well to take note of the appearance of certain weeds in his garden to discover if they are an indication of any special fault. There is a good deal of truth in the saying that weeds growing in a particular place are those beneficial to its fertility. In other words, weeds in a garden are rich in the particular minerals that the soil may lack.

However, let it be quite clear that these signs or indications of soil fertility or deficiencies, are general and not specific. Many other factors may influence the growth of certain weeds so these indications should always be taken as general.

A sudden crop of daisies may mean that the soil is lacking in lime. Daisies are rich in calcium and as they die and decompose they naturally enrich the soil with the calcium they have absorbed thus correcting the deficiency. Most weeds are very efficient at seeking out the minerals they require for their growth, in spite of an apparent scarcity, which is why many of them appear when the availability of certain nutrients in the soil decreases. Nature's way of correcting this deficiency is by the decomposition of these plants.

Daisy

One of the most reliable indicators of an acid clay soil in need of drainage is horsetail. Once this fault has been dealt with it will disappear only to reappear if the soil fertility falls and wet conditions return to the soil. Docks and sorrel usually indicate a badly drained and acid soil, and dandelions, with their long tap roots and ability to bring up available nutrients from the sub-soil, are often the first sign of some deficiency.

Indications

In nature, a plant only flourishes where both soil and its environment are favourable to its growth. Weeds accumulate the trace elements and minerals essential to the growing plant which tend to become exhausted in soil under constant cultivation.

There are many weeds that usually indicate an acid soil though most of them seem to prefer a clay to a light soil. Corn marigold and spurrey are found on light acid soils, but mayweed, plantain, dandelion and many others all indicate the heavier type of acid soil. We have mentioned that horsetail can indicate a badly drained and acid

Redshank

soil, and redshank is another indicator of water at a lower level, even though the surface may be dry.

If groundsel, one of the more common garden weeds, flowers when only about two inches high, and two or three crops appear in a season, then it is an indication of soil lacking in fertility. In a fertile soil, a well-grown groundsel plant will often reach a height of two feet, providing a rich source of material for the compost heap. Coltsfoot, with its penetrating rootstock, is one of the best sub-soilers and is very often found in heavy clay helping to aerate the soil. The perennial thistle and annual sow thistle are other sub-soilers which help with aeration and drainage.

Most gardeners will recognise the symptoms of nitrogen deficiency in their growing plants, a rather slow, restricted and spindly growth with a yellowing of leaves. But a strong vigorous growth of any of the leguminous weeds will also indicate a shortage of nitrogen. The bacteria living in the root nodules of these plants, which include clovers, medicks, vetches and wild pea, take up nitrogen from the air and store it in the nodules when it eventually passes into the plant.

This ability for nitrogen fixation is probably the reason for the rich green colour of these plants, and certainly a reason why they should be prized. If the tops are cut off and added to the compost heap, leaving the roots to decay in the ground, the soil will benefit from a free dressing of nitrogen.

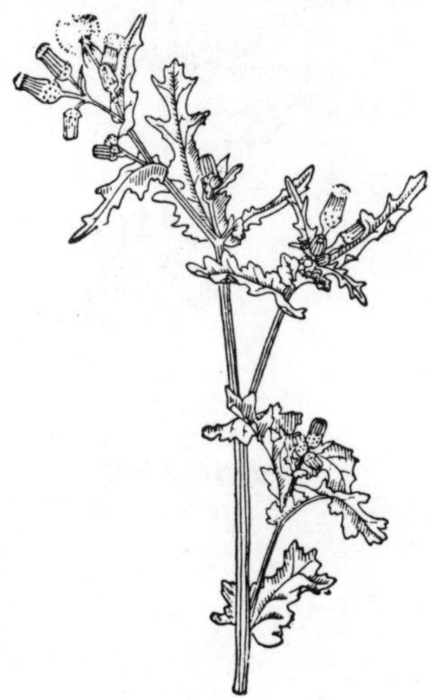

Groundsel

Tufted Vetch

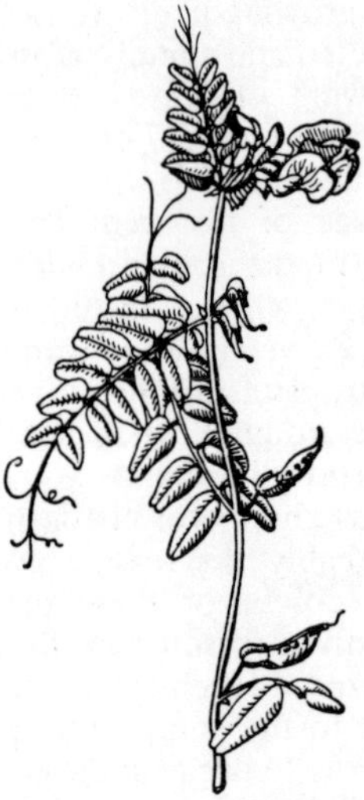

Bush Vetch

Clover

GENERAL USES

The Soil Association does not advocate the uncontrolled use of weeds in the garden, but it does recognise the fact that their controlled use can be beneficial. F.C. King, who was in charge of the famous gardens at Levens Hall for many years, wrote a number of books on compost gardening and was a staunch believer in the use and value of weeds. For example, he maintained that the best way to secure a good crop of sound onions was to allow the weeds to develop in the onion bed after about the first week in July. The growing weeds, by denying the onions a supply of nitrogen, improved their keeping qualities, and by digging in the weeds in the autumn, a supply of humus was made available for the next crop. This is an excellent example of the controlled use of weeds.

If you feel you *must* remove the weeds, don't burn them. Either add them to your compost heap or hoe them and let them lie on the ground to rot. By doing this you will return the organic matter to the soil as well as all the valuable trace elements and minerals the weeds contain and which are so essential to the growing plant. Burning destroys organic matter and unless you always make a point of returning the ash to the soil, you are constantly robbing the soil of the nutrients collected by the weeds. Weeds and compost go together and the regular use of properly made compost applied to the surface of the soil will build up soil fertility so that the plant requirements are met in full by the soil. If there is a particular reason why you do not wish the weeds to lie on the ground to rot, then put them on the compost heap. Weeds are usually *the only whole plant* added to the compost heap, other plants have either been harvested or have had part of them removed.

Obviously, weeds should not be encouraged where there are seedlings and young plants, but should be pulled and left to decay so that they will naturally enrich the soil. Once the plants have become established the weeds offer no competition and in many cases can be beneficial providing their growth is controlled. Their value lies in the proportion of weeds to plants, there is room for both. Again F.C. King gives many examples of the value of allowing weeds to remain with certain crops, instancing especially raspberries and strawberries. A good crop of weeds appeared to control and prevent the spread of red core disease and viruses in diseased strawberry plants specifically planted to test the result. Mr. King does not specify any particular weeds, but one must assume that couch grass, bind-weed and others of a similar nature would not be tolerated under these conditions.

Bindweed

In the case of the raspberries, a patch of coltsfoot was the problem. An attempt to remove the roots of this weed so damaged the roots of the raspberry canes that Mr. King's answer was to allow the roots to continue with their sub-soiling activities, but to remove all the top growth twice a year. This provided valuable material for the compost bin and the soil benefited to the obvious advantage of the raspberries.

During the late summer and autumn weeds should be allowed to flourish in order to prevent the loss of valuable plant food at a time when the crops are ripeneing and do not require much nourishment, and also after the crops have been removed. Nature never leaves the ground uncovered, and during the winter, weeds will give some protection from the weather, the roots helping with aeration and drainage and providing food for the bacteria. This bacterial activity will help to build up a store of available plant food for the spring and will provide warmth in the soil for spring planting. Always remember, where ground cover has been allowed to remain and flourish there will be a build up of soil fertility. The dead and rotting remains of top growth and roots provide nutrients and humus, and the roots also help with the aeration and drainage.

A return of well made compost to build up soil fertility often results in the disappearance of certain weeds, showing that the balance of nutrients is beginning to right itself. It is always interesting to notice how the return of this organic matter, by increasing the fertility of the soil, effectively alters the species of the weeds in the area. Those that indicate poor and acid soil will, as the fertility increases, be replaced by others, perhaps showing a mineral deficiency, until, when a high standard of fertility is eventually reached, they too will be replaced by weeds showing by their growth and type, this high degree of fertility. The observant gardener will notice that during this period the first to appear in a bad soil will be the deep rooters which will eventually give way to the more shallow rooted plants, showing an increasingly fertile soil. Again, one would caution that these indications are general and not specific.

An extract from the Soil Association Journal, 'Mother Earth' is worth quoting here:

> "In an article on 'Some aspects of the cultural control of weeds', in the 'Journal of the Royal Horticultural Society' for May 1952, Robert Adams referred to an address by Mr. F.A. Secrett, the well-known market grower, to the Royal Society of Arts. When he first started his farm at Milford, Mr. Secrett found the chief weeds to be spurrey and, among the perennials, the deep rooted horsetail. Immediately, he applied large quantities of organic matter to increase the soil fertility and under his cropping plan, these weeds gradually died out, to be replaced by groundsel, annual meadow grass and chickweed. Later, as a result of the sound cultural methods employed, a very high standard of fertility was reached and the weeds mentioned were in turn replaced by the dwarf annual nettle. By these purely cultural practices Mr. Secrett has finally eliminated

the perennial horsetail and now has the comparatively easy task of controlling a shallow rooted annual."

Many of you may wonder how the application of good compost will help with the aeration and drainage of the soil. This will encourage the earthworm population whose long tunnels drain off surplus water and admit air to the sub-soil. Worms are humus makers and the best sub-soilers in the business, but they cannot live in an acid soil.

Some years ago, one of our members, Miss K. Keswick, wrote a very interesting account, also published in the Soil Association Journal, describing her attempts to turn a field into a garden. She had always intended to do as little digging as possible, but circumstances occurred which prevented her from even hoeing or weeding, except very occasionally. Each year saw a change in the number and variety of weeds appearing. Thistles and charlock grew in profusion, then plantain, docks and buttercups appeared, all playing their part in penetrating deep into the sub-soil helping with aeration and breaking up the heavy clay. As her garden progressed so did the building of her house, and when the roof was completed, rainwater flowed from the gutters into soakaways and then into the centre of the garden which became a swamp. Watercress grew, buttercups multiplied, showing sourness, and moss appeared. Obviously, land drains would have to be laid, and when this work had been completed, the watercress disappeared to be replaced by fat hen – usually an indication of good organic matter – followed by mayweed and pimpernel. The following year a mass of chickweed appeared and at this stage, an old gardener remarked that next year she would have groundsel, and then her garden would be made!

Fat Hen

To complete this chapter on the general used of weeds, bear in mind that every effort should be made to hoe the weeds before they seed, especially if they are to be left on the ground to decompose in situ. This is not so vital if the weeds are to be added to the compost heap, the important point here is that the compost should be made properly in order to get rid of any seeds it might contain. The Soil Association booklet "Garden Compost" covers the practical do's and don'ts of garden compost making and explains clearly how this can be achieved.

Examples

Some examples of the more common weeds may help with general indications. They are not confined to those found in urban gardens but include weeds that find their way, one way or another, into the country garden.

Coltsfoot, creeping buttercup, creeping bent grass, dandelion, plantain, daisy, and the broad leaved dock are all to to be found on heavy soils. Bracken, sheep's sorrel, common sorrel, plantain, spurrey, corn marigold, sow thistle, scentless mayweed, and the daisy can all show acidity or the absence of lime. Campanula, salad burnet, scarlet pimpernel, and bladder campion are generally found on calcareous soils; while spurrey, corn marigold, sheep's sorrel, cornflower, small bindweed, small nettle, shepherd's purse, white campion and most of the shallow rooted weeds prefer a light or sandy soil. Weeds found on damp, and in many cases, badly drained soils, include horsetail, redshank, silverweed, creeping buttercup, and some mosses. Many of the weeds appear in more than one category. For instance, the creeping buttercup is usually found on heavy land but can also indicate a need for drainage. The keen gardener will learn much about his soil by observing the prevalence of certain weeds, their type of growth – is it strong or poor – and whether the addition of organic matter to the soil is helping to control them.

Hoary Plantain

Stinking Mayweed

SPECIFIC USES

To many people this will be the most important chapter of all, because weeds can be used effectively as a natural liquid fertiliser, insect repellent, foliar spray, and as a preventative against fungal attacks.

Natural liquid fertiliser

We know that weeds have the capacity to extract minerals from the sub-soil and if they are gathered fresh in the morning and steeped in a bucket of rainwater for a couple of weeks, the resulting liquid will make an excellent plant manure, returning all the nutrients collected by the weeds, to the growing plants. Use a good mixture of weeds and any amount you like. The liquid can be used directly on the foliage or around the roots. If you use a watering can, strain the liquid through a piece of cheese cloth so that the rose will not get clogged up with pieces of vegetation.

Stinging Nettle

Heading the list of beneficial weeds must surely be the stinging nettle, for its uses are legion. It is the most companionable, or symbiotic plant that exists and it is amazing how much time, energy and money the average gardener will expend on trying to rid himself of this most valuable plant.

It has the ability to stimulate the growth of other plants growing nearby, perhaps because of its high nitrogen content, and to make them more resistant to disease. More than one gardener has found that the yield and quality of soft fruits have been improved if nettles are allowed to flourish near the plants.

As a companion plant to herbs it seems to increase the essential oils in aromatic herbs.

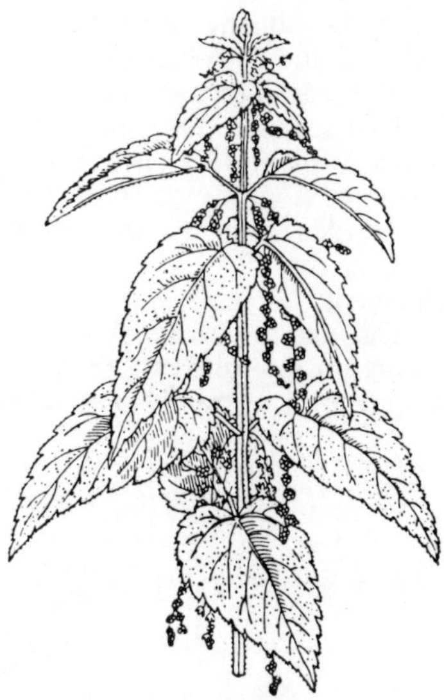

Stinging Nettle

It breaks down into almost perfect humus and is invaluable as a mulch between rows of vegetables providing both humus and nutrients.

Because of its high mineral content it is an essential addition to the compost heap. When added freshly cut, or only slightly wilted, it has the effect of raising the temperature of the heap almost at once, encouraging the nitrogen bacteria in the breaking down and decomposition of vegetable matter, especially its own tough fibrous stems and roots.

Nettles can be left to soak in rainwater for two to three weeks until they disintegrate. The rather smelly liquid thus produced makes a very effective plant food and is especially good for tomatoes. Again, the amount of nettles to be used is immaterial. In many cases the supply will dictate this. However, the more you use, the stronger will be the plant food, and bear in mind that there is more nitrogen in fresh, young nettles during the spring and summer than there is in the old plants during the autumn. This liquid can be used as an insect repellent and as a foliar feed. Or an infusion can be made by covering a handful of the fresh nettles with a pint of water, bring to the boil, remove from heat and cover while cooling. Strain and dilute with four parts of water to one of liquid. This can be used as a spray against mildew, blackfly, aphis and plant lice in the greenhouse or outside. When used as a foliar feed a dessertspoonful of liquid soap can be added to the spray to help it to stick to the leaves, but this should be omitted if the liquid is to be used as a fertiliser.

As the stinging nettle plant is the only food of the caterpillars of the Peacock and Small Tortoiseshell butterflies they should be cultivated for this reason too.

That seems to be a pretty impressive list of uses. The stinging nettle is also a plant very rich in protein, vitamins and minerals, especially nitrogen, iron and calcium, and not only is it of tremendous value in the garden, but it has long been recognised for its medicinal and culinary uses.

Horsetail

This plant, with its almost endless tap root, has a great capacity for taking up soil nutrients according to their availability. As well as being an indicator of poorly drained soil, it secretes great quantities of cobalt

and calcium and is particularly rich in silica. For this reason an infusion made from this plant can be used as a preventative against fungal diseases, but should not be regarded as a cure for the disease. It has been found helpful against black spot on roses and mildew and other fungi on grapevines, vegetables, roses and fruit trees. It can also be used to prevent mint rust and the rust that attacks mallows such as hollyhocks. It should be borne in mind, however, that this is not an infallible remedy, but one that is certainly worth experimenting with.

The branching shoots of the horsetail, rather like little pine trees, are the richest in silica and can be used either fresh or dried. Using fresh, put two good handfuls in a pan and cover with water, bring to the boil and boil for twenty minutes. Cover and cool overnight. Strain and dilute with two parts of water to one of liquid. Spray only on plants which have their true leaves developed and dilute the liquid more each time it is used in the same place. It is comforting to know that some use can be made of this plant in the garden for it seems to be the bugbear of so many gardeners.

Chamomile

I refer here to the true wild chamomile – chamomilla matricaria – and not to the little prostrate chamomile – anthemis nobilis – used to make lawns. The wild chamomile, sometimes called 'German' chamomile, is a lime lover and is often referred to as the plants' physician. It appears to be medicinally helpful to plants growing nearby and sickly plants will very often perk up if chamomile is planted beside it. Chamomile tea can be used to prevent 'damping off' of seedlings. Soak the blossoms in cold water for some hours and use the solution to water the seed boxes, use the same solution to control mildew on cucumber seedlings. Because of its high calcium content it is a most useful plant in the compost heap and helps to keep it 'sweet'.

Chamomile

SOURCES OF MINERALS AND TRACE ELEMENTS

Bulbous Buttercup

cobalt
copper
iron
manganese
phosphorus
potassium

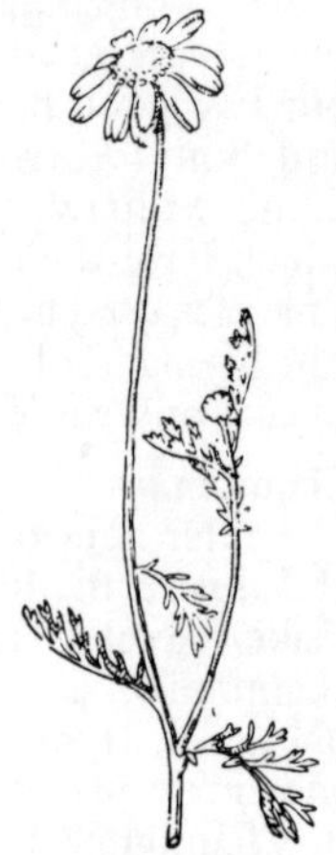

Chamomile

calcium
potassium

useful plant in the compost heap.

Broad Leaved Dock

phosphorus
potassium

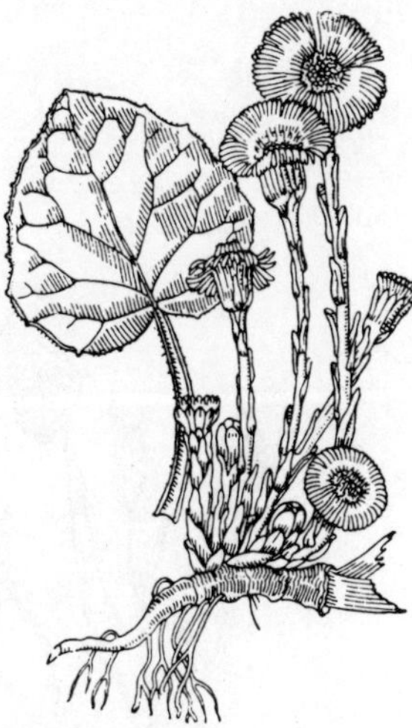

Coltsfoot

calcium
copper
iron
magnesium
potassium
sulphur

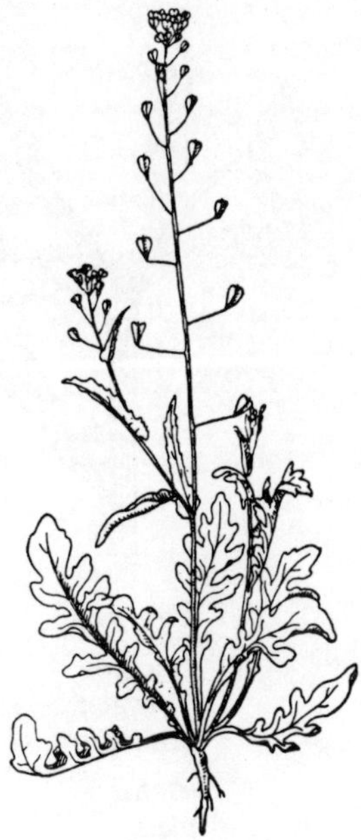

Shepherd's Purse

calcium

Chickweed

manganese
potassium

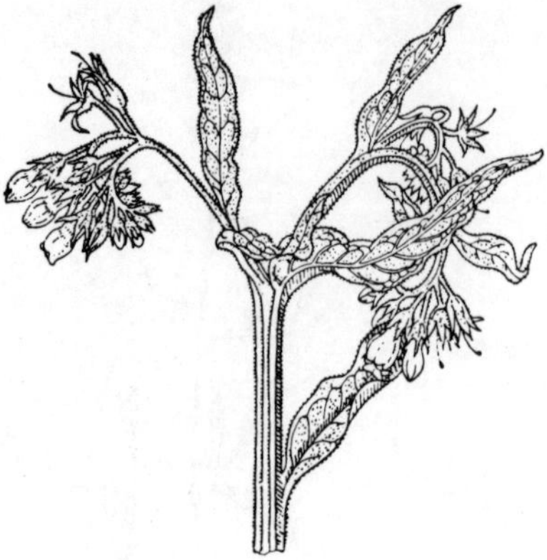

Comfrey

calcium
nitrogen
potassium

useful plant in the compost heap.

Creeping Thistle

calcium
copper
iron
potassium

Daisy

calcium
magnesium

Dandelion

calcium
copper
iron

Fat Hen

calcium
iron

Groundsel

iron

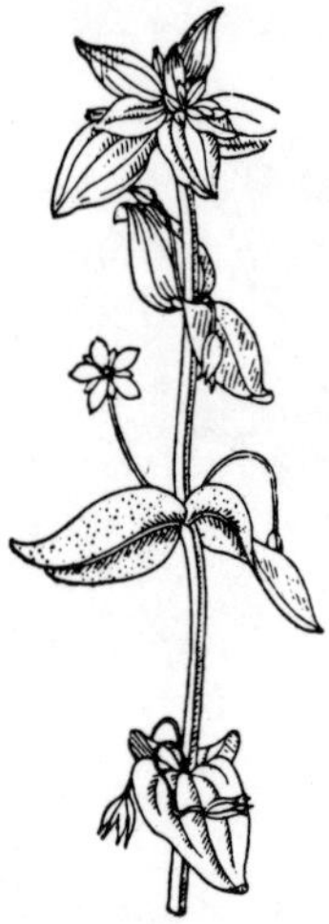

Pimpernel
calcium

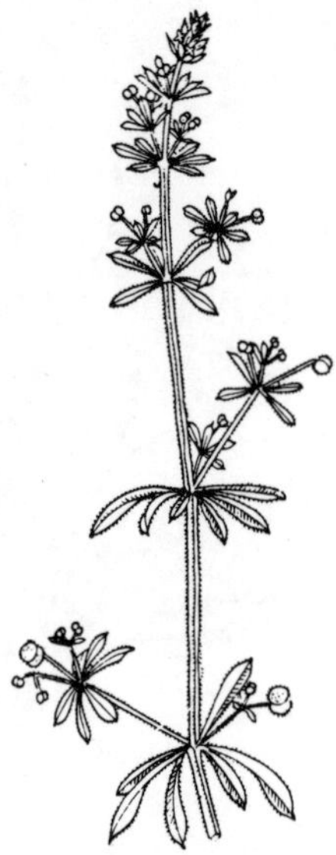

Cleavers
calcium

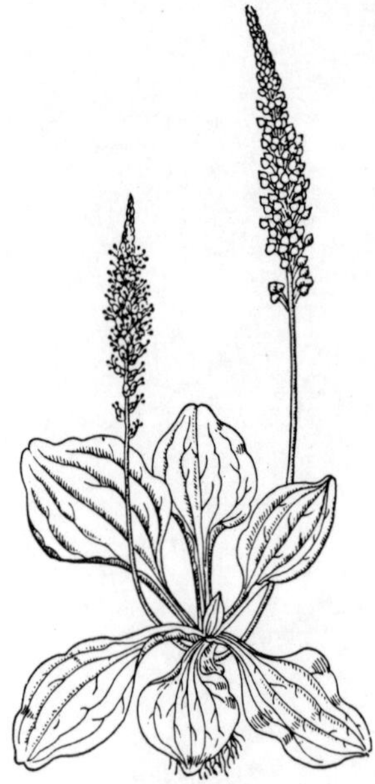

Great Plantain
calcium
copper
iron
potassium
silica

Horsetail
calcium
cobalt
iron
magnesium
silica

Silverweed

iron
calcium
magnesium

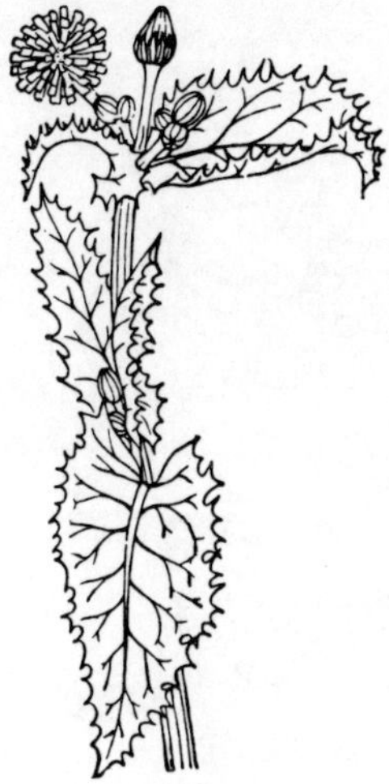

Sow Thistle

copper
potassium

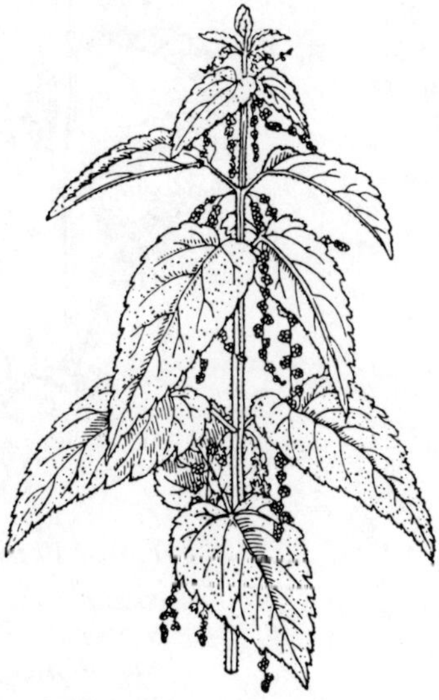

Stinging Nettle

calcium
copper
iron
nitrogen
potassium
excellent in the compost heap

Sun Spurge

boron

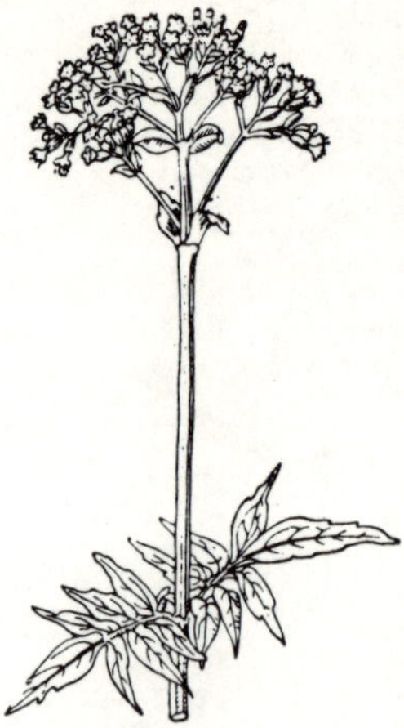

Valerian

rich in
silica
valuable
plant in
compost
heap

Yarrow

copper
nitrogen
phosphorus
potassium

valuable plant in
compost heap.

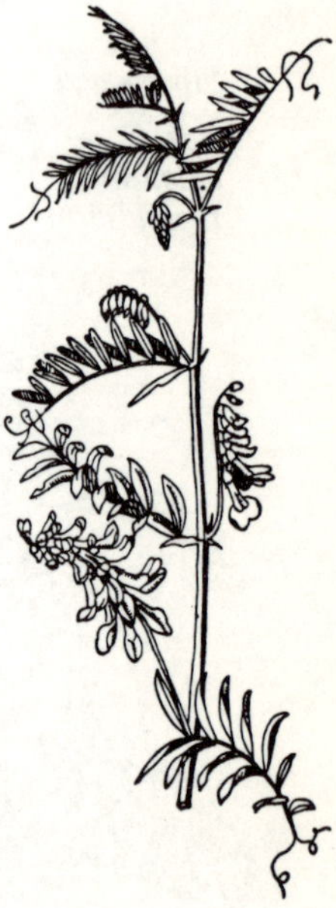

Tufted Vetch

cobalt
copper
phosphorus
potassium

THE SOIL ASSOCIATION

President: Dr. E.F. Schumacher, C.B.E.

A world wide charity, founded in 1946, to promote a fuller understanding of the vital relationship between soil, plant, animal and man. The Association believes that these are parts of one whole, and that nutrition derived from a balanced living soil is the greatest single contribution to health (wholeness). For this reason it encourages an ecological approach and offers organic husbandry as a viable alternative to modern intensive methods.

Current activities of the Association are:

To promote organic husbandry as a viable alternative to modern intensive methods.

To assist farmers, horticulturists and gardeners who wish to adopt organic methods.

To conduct courses on the principles and practices of organic husbandry as part of its expanding educational programme.

To publish basic information on organic practice to support this programme.

To promote research and arrange field trials in organic methods.

To take part in conferences, exhibitions, agricultural and horticultural shows in order to demonstrate the advantages of organic methods.

To provide qualified speakers.

To publish a Quarterly Review free to members and to provide a comprehensive Bookshop service.

The Association is supported by active Groups throughout the United Kingdom.

The Association has published a code governing methods of organic production to the highest quality. The Association has its own registered Trade Mark which it issues on licence to growers who conform strictly to this code. This Trade Mark on any produce for sale is a sign of its high organic quality.

JOIN THE SOIL ASSOCIATION NOW

Use the Form overleaf

APPLICATION FORM

Annual Subscription £5.00

VW

Please complete this form and send it to:
THE SOIL ASSOCIATION,
Walnut Tree Manor,
Haughley,
Stowmarket, SUFFOLK IP14 3RS.

Name ....

Address ....

....

Occupation ....

Age Under 25 : 25 to 45 : 45 to 65

We will put you in contact with your local Group. Information on your occupation will enable us to put you in touch with like-minded members.

The Association is a Charity and welcomes donations to assist it in its work.

Payment can be made as you like, cheque, Postal Order or cash. A Banker's Order is easy and saves the Association much clerical work and postage. A covenanted subscription costs you no more, but in real terms benefits the Association by some 45%.

BANKERS ORDER FORM

To the ....Bank

.... Branch

Please pay to Barclays Bank (20-82-75) Stowmarket, Suffolk for the credit of the Soil Association (10811963) on receipt.

of this order and on the .... day of ....

in each following year the sum of £ ....

Signature .... Date ....

Address ....

....